IRIDOLOGY

IRIS ANALYSIS

When

YOUR EYES ARE THE LIVE CAMERAS TO YOUR BODY

Foreword

When I am not in my waters holding Water-Chi sessions, in my spare time, I love to study different fields of interest to me. Iridology has been one of them. This white paper is made available to you for the sole purpose of sharing my findings about Iridology. I am a life learner and I will always be… Over the last twenty years, I have studied several subjects for my personal interest (phlebotomy, laboratory analysis at ROP, Regional Occupational Program in Northern California). There, I volunteered my free time at a local AIDS clinic to draw blood for HIV tests and at the Coroner's Office. I've done personal research about holistic healing and graphology (handwriting analysis).

Iridology is a fascinating subject as it makes us question modern medicine which purpose is to treat symptoms. Iridology has a holistic approach of health treating the body. An iridology analysis determines where the health predicament or problem comes from, which is not necessary where the symptom is located.

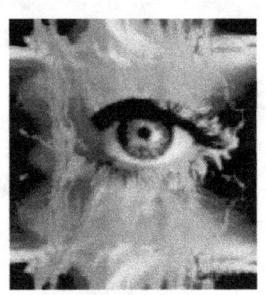

Iridology - Herbal Medicines

This white paper is solely intended to enlighten readers about the concept of Iridology and its interpretations and does not intend to deliver diagnosis, or medical advices. If you use the information contained herein without the advice of your physician, you are prescribing for yourself (which is your constitutional right), I assume no responsibility.

There are profuse amounts of herbal medicines/remedies that are commonly used to help treat health conditions. It is best to consult with a Naturopath or an Homeopath physician. Mixing potent tinctures or herbs may be harmful without receiving professional advice.

"The Doctor of the future will give no medicine, but will interest his patients in the care of the human frame and the cause and prevention of disease."

Thomas Edison

Background

Thousands of years ago, our ancestors from different continents relied on examining the eyes to make diagnostics. Long before modern medicines, Europeans, Greeks, American Indians, Egyptians, Chinese relied on herbal remedies and physical understanding… Hyppocrates, the legendary Greek physician and teach in medicine, born 460 BC, established a link between signs in the eyes and changes in the body. Finally, in the mid-1800's, a bright Physician, Dr. Ignatz Peczely furthered his knowledge into this phenomenon we now call Ophthalmic-Somatic Analysis (Iridology).

Your eyes, more precisely, your irises, are the cameras to your internal body. With a magnifying glass, look at it carefully… It is like an open book that reads into your past, present and future health.

Iridology is the science that uses the iris of the eye to diagnose and monitor tissue changes that are occurring or have occurred within the body. It is possible to detect under-active organs or tissues and overactive organs and tissues.

This white paper is meant to be user-friendly and it should help you identify healthy or unhealthy signs as they occur by monitoring the eyes with a handheld magnifying glass or a close up picture of the eyes. You will find a few herbal healing considerations listed, which may be applicable to some diagnosis. The potent healing of nature through herbal formulas help balance the body. Therefore, it is best to buy ready-made products at your local natural food and drugs store.

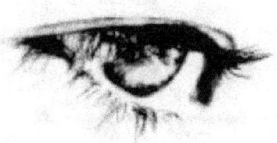

ORGAN SIGNS

- The following shapes of color can be found in the iris. They pinpoint the different organs and the zones in the iris where a problem may have occurred or is occurring:

- LEAF LACUNA: located in thoracic region of the iris, and is common to lung or heart weakness

- OPEN LACUNA: Located anywhere in the iris, shows an area of weakness.

- CRYPT: Normally seen in the glandular zone. When several are seen and they are dark, they indicate clinical manifestations of glandular problems (diabetes mellitus, hypoglycemia, etc.)

- OPEN CRYPT: It can be seen anywhere, with comparative value of CRYPT.

- KIDNEY MEDUSA: Inherited kidney weakness, check zone.

- LIVER STAKE: Most often seen in the liver zone. Shows inherited weakness. Tar black color signify area susceptible to becoming cancerous.

- HONEYCOMB: Indicates functional metabolism occurring at that location. Can diminish as problem is corrected.

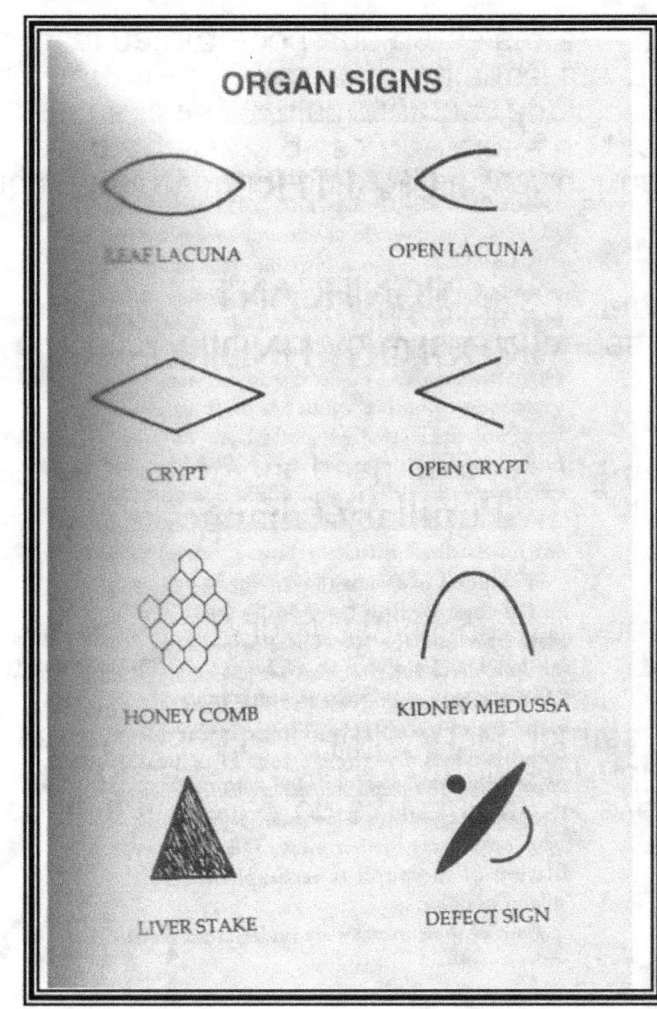

Iridology Chart

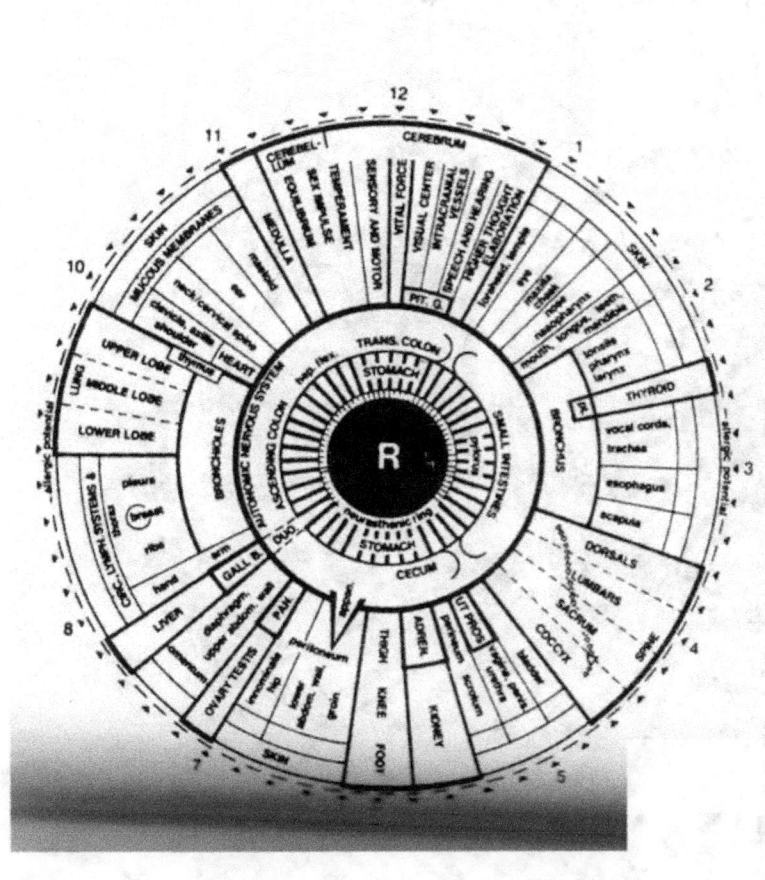

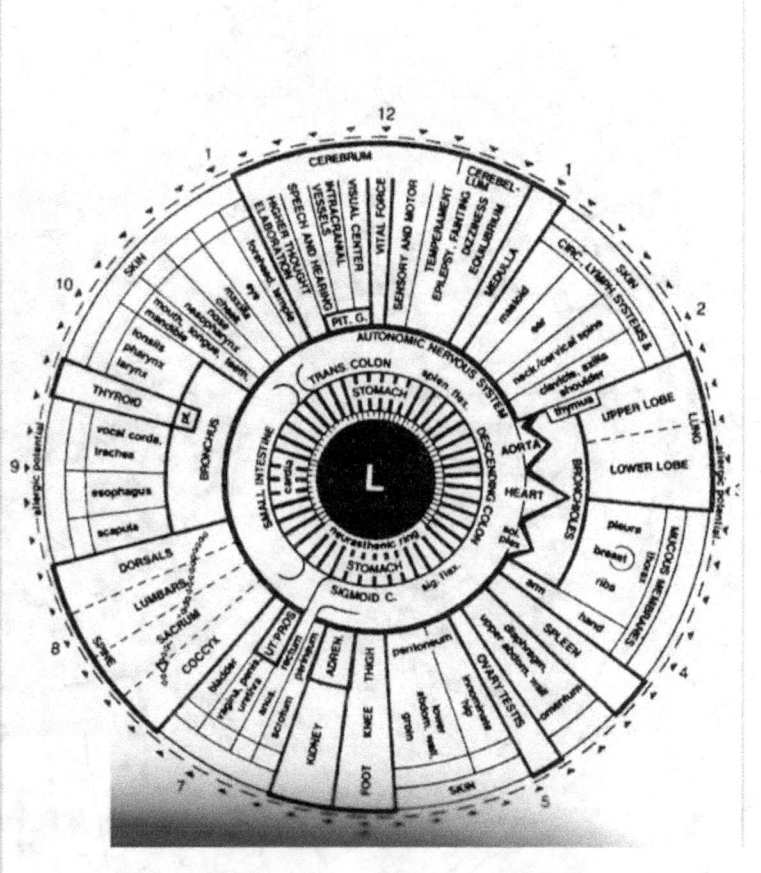

Right Eye

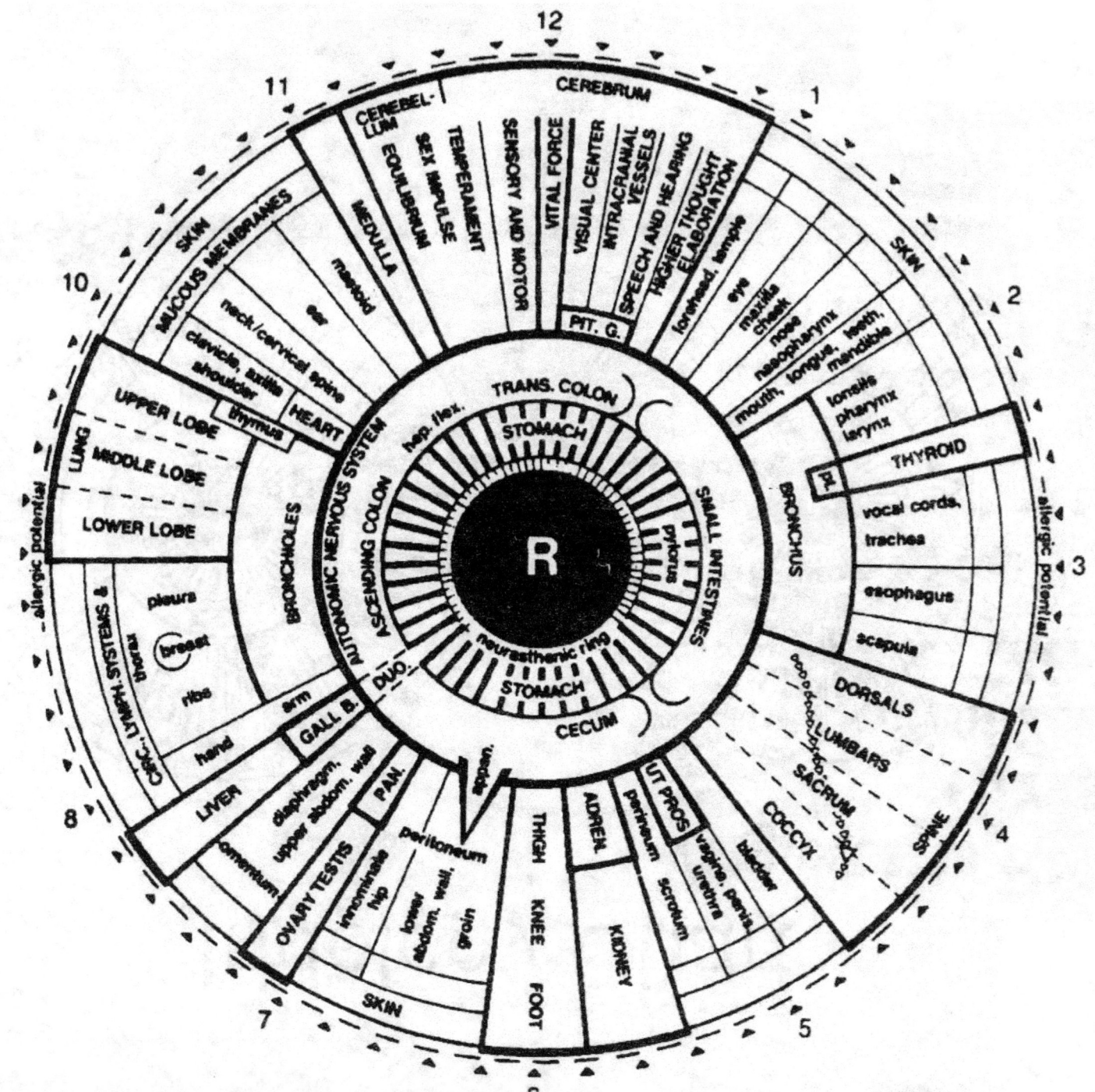

Left Eye

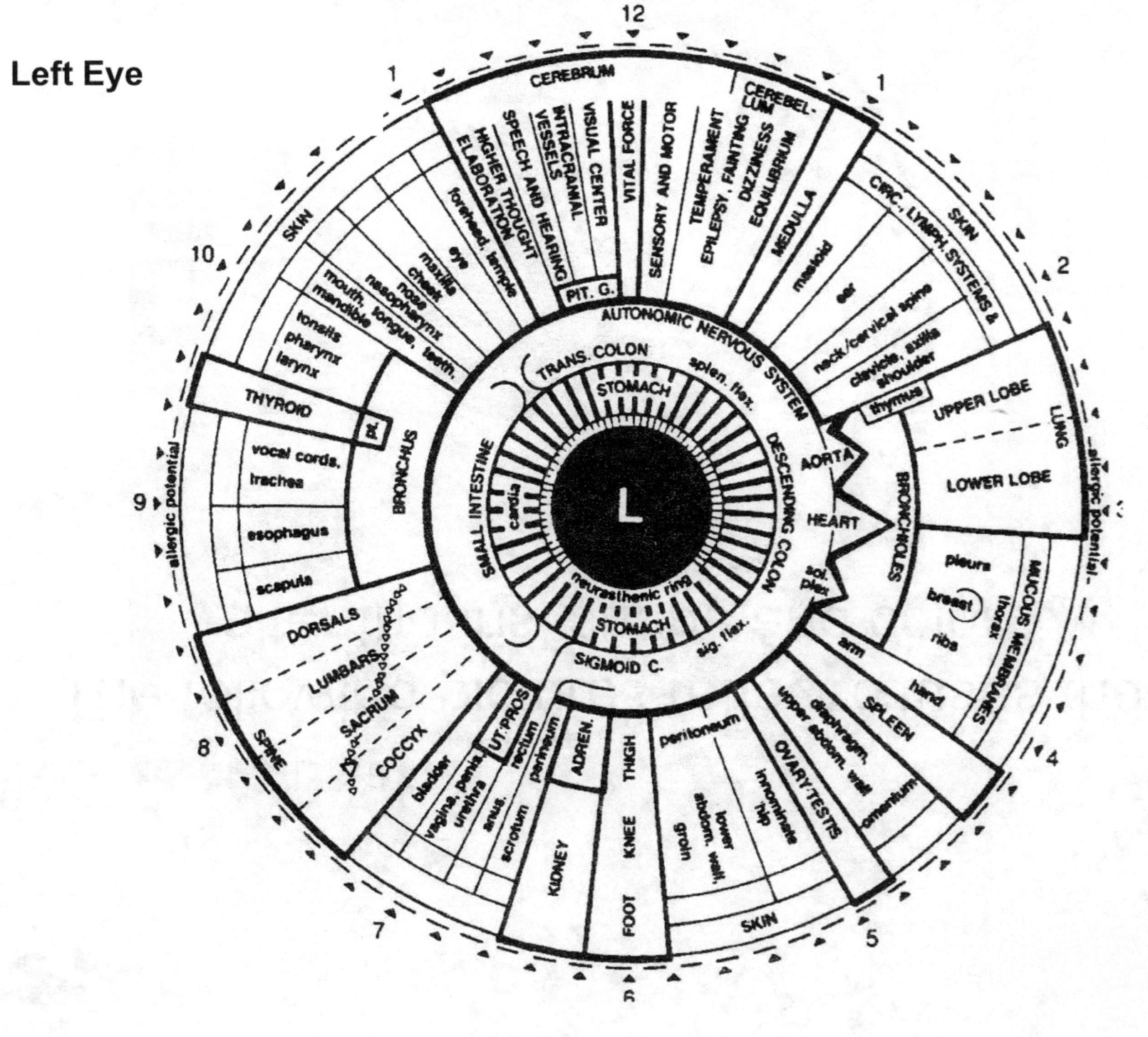

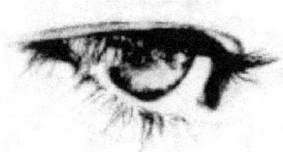

The following pictures reflect causes that alter the iris' shape and color

STOMACH HYPOACTIVE

- Cause: malnutrition

- Iris: Dark ring around the pupil in the stomach zone of the iris

- Symptoms: easy nausea, belching, burping after meals, general feeling of weakness

- Herbal suggestions: Papaya, mint

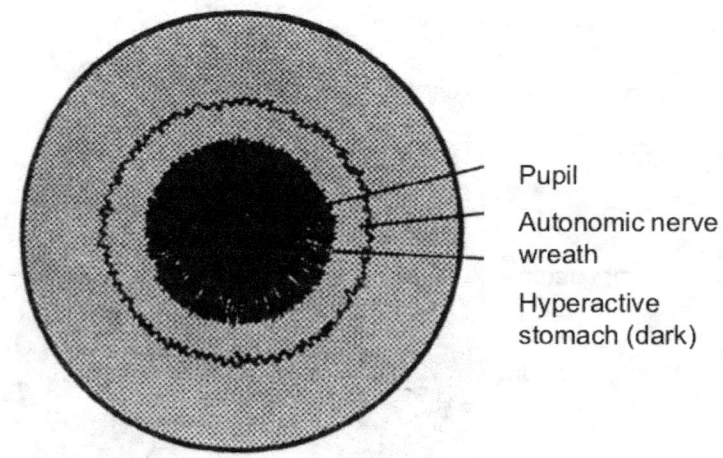

Pupil

Autonomic nerve wreath

Hyperactive stomach (dark)

STOMACH HYPERACTIVE

- Cause: Coffee, alcohol, tobacco

- Iris: White ring around the pupil, the dysfunction is one of overactive gastric/stomach Symptoms: spit-up excessive digestive acid, may lead to ulcer

- Herbal suggestions: One to two tablespoons of aloe vera liquid 15 minutes before meals. Capsicum can heal a bleeding ulcer almost completely within 48 hours

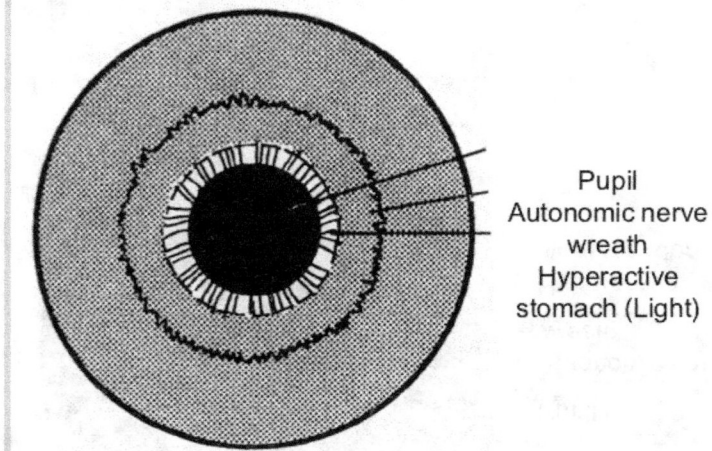

Pupil
Autonomic nerve wreath
Hyperactive stomach (Light)

RADII SOLARIS

- Cause: Presence of parasites

- Symptoms: sun rays like around the pupil. Irritation of the nervous system

- Herbal suggestions: Ginger, Fennel, turmeric root, garlic, pumpkin

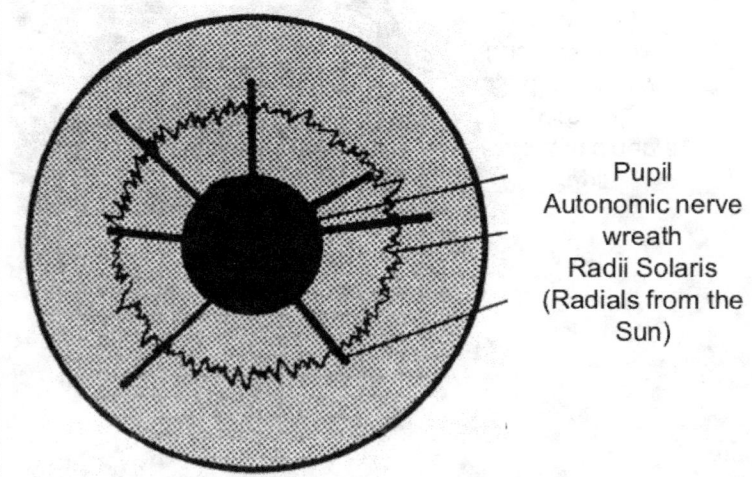

Pupil
Autonomic nerve wreath
Radii Solaris
(Radials from the Sun)

RADIAL FURROWS

- Cause: Increased Toxic Material

- Symptoms: Sunray burst around the autonomic nerve wreath. Weakness of the abdominal muscles. Lupus, Erythematosus, Multiple Sclerosis and many more

- Natural Healers: Bowel cleansing and Ginger, Fennel, turmeric root, garlic, pumpkin, black walnut, spearmint. Exercise (mini-trampoline)

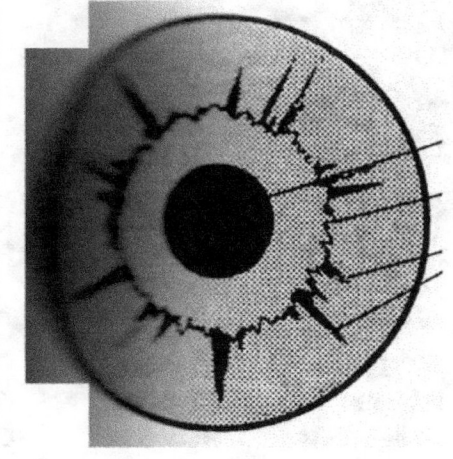

Pupil
Autonomic nerve wreath
Radial Furrows
(increased toxic materials)

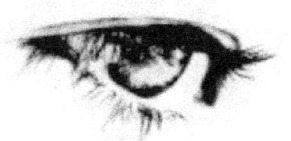

DIVERTICULI

- Cause: Waste material lodged in bowel wall. May cause colon cancer

- Symptoms: Small pockets like around the pupil. Constipation, diarrhea, stomach aches

- Natural Healers: Acidophilus, Aloe Vera Juice, Chamomile flowers, Rhubarb root, ginger

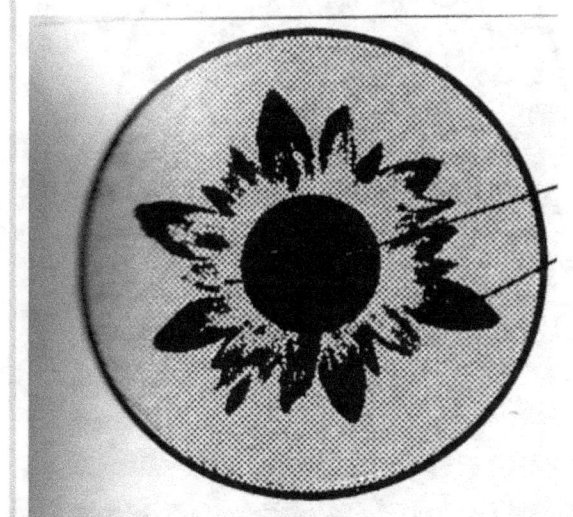

Pupil

Diverticuli

PROLAPSED COLON

- Cause: Very weak musculature of the intestinal wall

- Symptoms: Collapsed shape on top of the Autonomic nerve wreath. Constipation, prostate, uterus, bladder physical pressure, lumbar spine pain

- Natural Healers: Exercise, massage. Apple pectin, charcoal, ginger root, rose hips

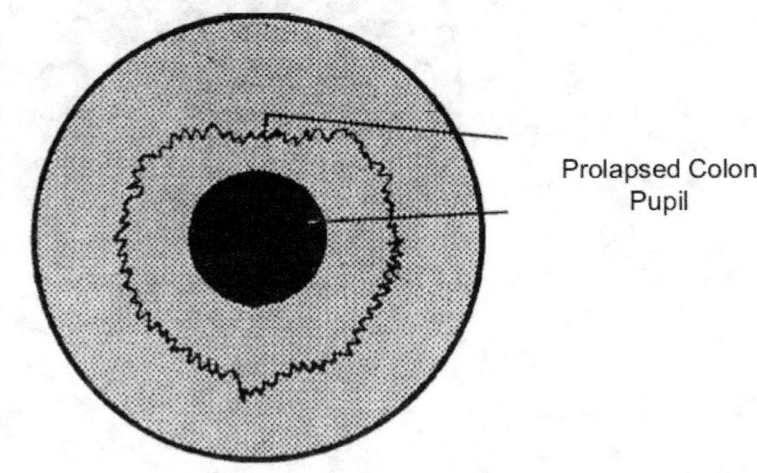

Prolapsed Colon
Pupil

HYPERACIDIC TISSUE

- Cause: Carbohydrates/sugars, coffee, tea, alcohol, citric acid (i.e. lemons)

- Symptoms: Multitude of thin white rays around the pupil. Acid stomach. It depletes natural sodium from the joints, thus creates calcium build up in the joints. It is the underlying cause of much of the arthritis and rheumatoid arthritis

- Natural Healers: East fish or fowl, celery, parsley, kelp. Aloe Vera, rose hips, Vitamin C

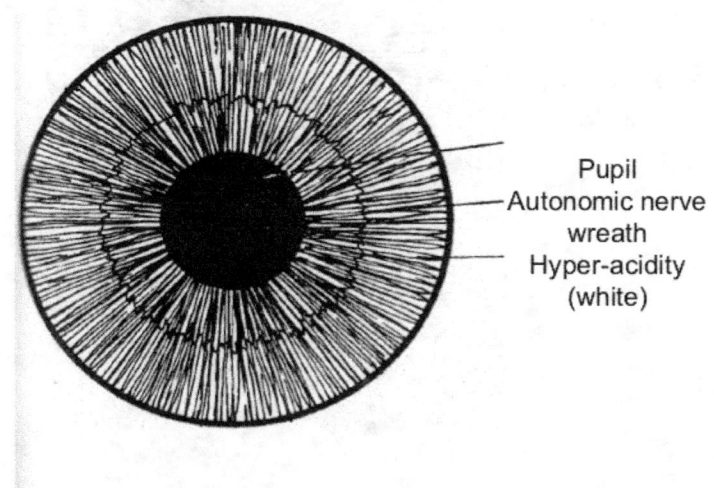

Pupil
Autonomic nerve wreath
Hyper-acidity (white)

HYPERALKALINE TISSUE

- Cause: Anta-acids such as Rolaids, Tums, white flour products, baking soda.. The inorganic sodium displaces potassium in the system, needed in the muscles and in the heart muscles

- Symptoms: Thicker whiter white rays all around the iris. Potassium gets stored in the intestinal tract and gets depleted under a condition of diarrhea

- Natural Healers: Intake of natural sodium, such as parsley, sage, celery. High potassium intake, such as bananas (vine ripe). Chamomile, dandelion

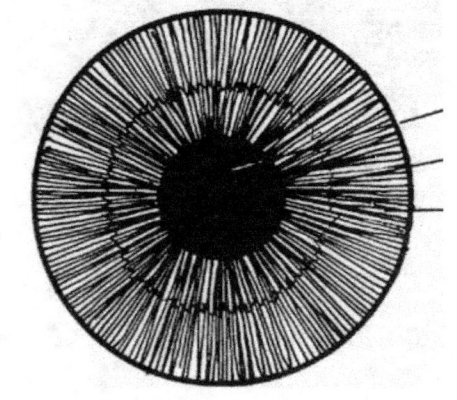

Pupil
Autonomic nerve wreath
Hyperalkaline Tissues (white)

PSYCHOSOMATIC STRESS RINGS

- Cause: Poor absorption of calcium, abnormal gastro-intestinal activity

- Symptoms: Stress lines on the iris. Kidney stones, plaquing (which always leads to calcium deficiency), depression, irritability

- Natural Healers: Alfalfa, high B complex, fennel seeds, chamomile flowers, St. john's wort, valerian root, passion flowers

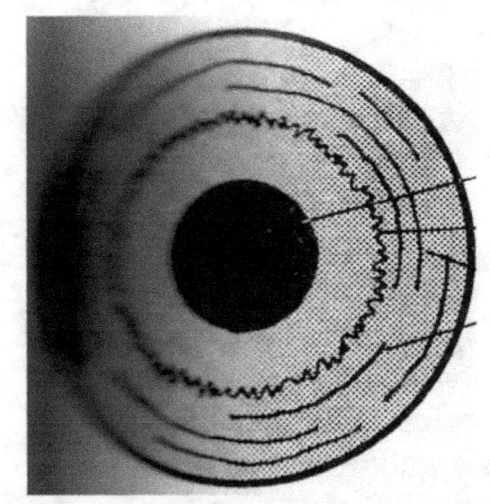

Pupil
Autonomic nerve wreath

Psychological stress lines

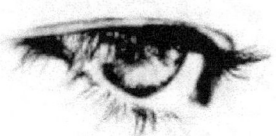

PATHOLOGICAL POLYCHROMIA

- Cause: Organ cell damage.
- Symptoms: May appear on the iris like freckles

- Symptoms: Freckles of different colors such as

- Dark brown: liver
- Reddish brown: Blood
- Orange: Pancreas
- Yellow: Kidneys

- The organ that causes the damage is pointed by the freckle.

- Natural Healers: Vary with each organ the freckles indicate

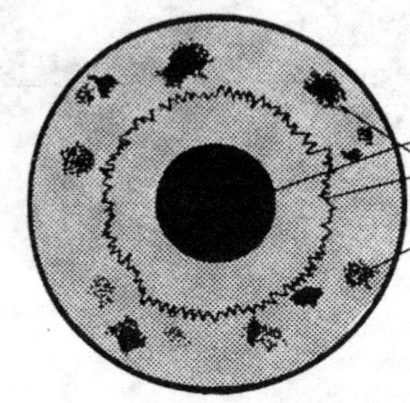

Pupil
Autonomic nerve wreath
Pathological polychromia

SECTORAL HETECHROMIA

- Cause: Drug, heavy metal or inorganic substance absorption

- Symptoms: a large area of a different color seen in the eye

- Natural Healers: Identify which (if) metal is involved. Drink only purified water, purify blood with i.e., red clover, dandelion, yarrow… Kelp and garlic are also effective in detoxifying the body of certain metals

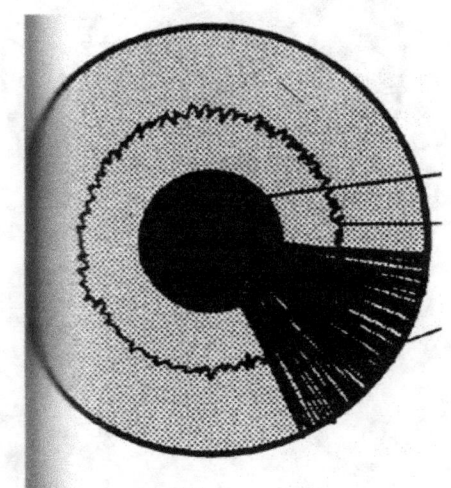

Pupil
Autonomic nerve wreath

Sectoral Hetechromia

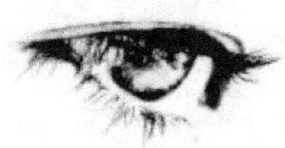

LYMPHATIC TOPHI

- Cause: Unknown

- Symptoms: Cotton ball shapes (various colors). Weak ability to protect and defend the system and weak ability to absorb fat soluble (D,A,K, and E). Weak ability to help purify the blood (plasma proteins back into the blood)

- Natural Healers: Blood purifiers (rose hips, Pau D'Arco, Echinacea, etc.). Exercise that promotes sweating

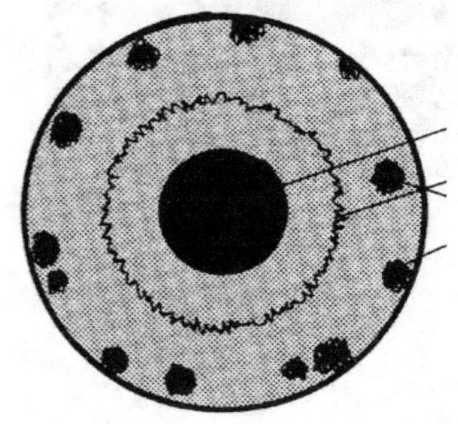

Pupil
Autonomic nerve wreath
Lymphatic Tophi
(cotton balls shape)

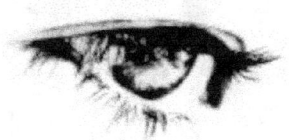

ARCUS SENILIS

- Cause: Alzheimer, senility. A partial blockage in the arteries of the head and neck primarily. Vessels begin to plaque from triglycerides, inorganic sodium, etc. forming a blockage of the arteries

- Symptoms: A white hat covering approximately the upper one third of the iris. White area is on the edge of the iris

- Natural Healers: Ginkgo and gotu kola. Vitamin E, A B complex. Slant board, rebounding types of exercising to increase blood flow to the brain

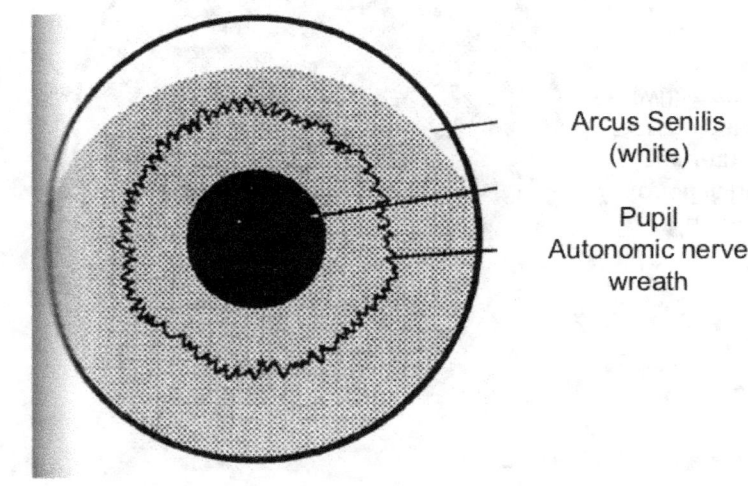

Arcus Senilis
(white)

Pupil
Autonomic nerve
wreath

HYPERCHOLESTEROSIS

- Cause: Circulatory blockage anywhere in the body due to (i.e.) salt, artificial sweeteners, etc. May lead to cardiac arrest when functional demands of blood is not met

- Symptoms: White ring around the iris. Brilliance of the reflected light from this area and the thickness of the ring vary with severity of condition

- Natural Healers: Eliminate salt. Clean liver with blood cleanser red clover, rose hips, horseradish. Add Omega 3 EPA to diet

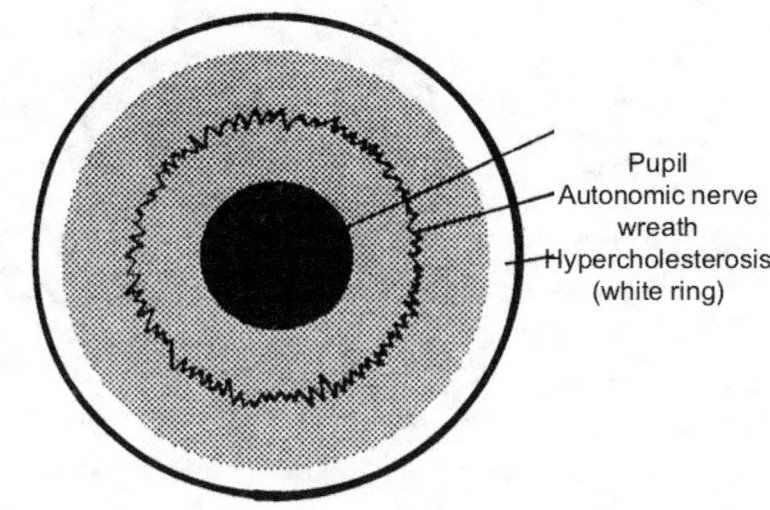

Pupil
Autonomic nerve wreath
Hypercholesterosis (white ring)

TOXIC CIRCULATORY SYSTEM

- Cause: Blood has excessive toxins seeping into it from both the large and small intestine

- Symptoms: a dark ring around the iris

- Natural Healers: Red clover, burdock, yarrow, etc.. Only one single herb boost should be used for burdock or Echinacea

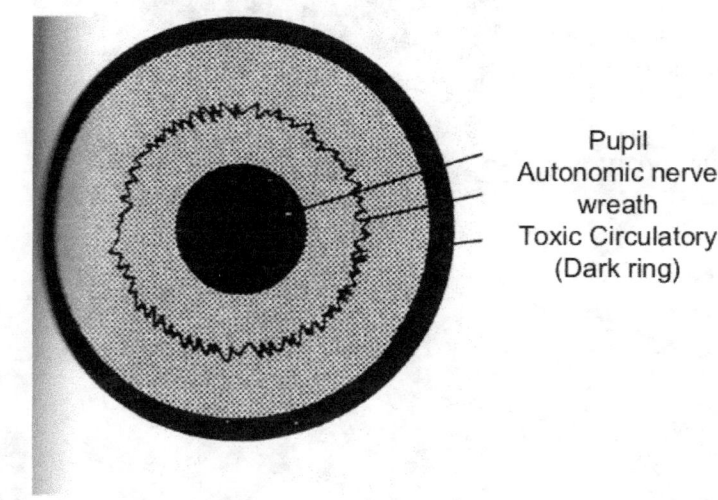

Pupil
Autonomic nerve wreath
Toxic Circulatory (Dark ring)

SCURF RIM

- Cause: Many. Kidneys are not eliminating its acids (uric acids), B-complex and minerals deficiency. Poor air supply through the skin

- Symptoms: Iris looks a bit cloudy and scurfy around rim. Gout (acute inflammations, especially toes) Skin sores and eruptions

- Natural Healers: Herb compositions and regulated diets

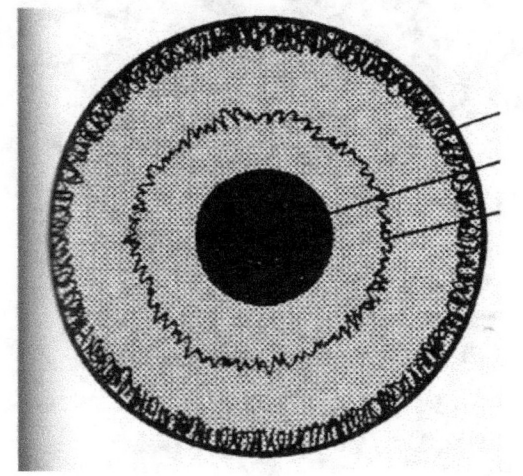

Scurf Rim
Pupil

Autonomic nerve wreath

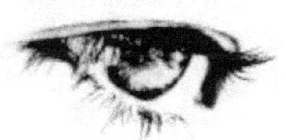

HEALING CRISIS

- Cause: The body is actively fighting a condition (fever, pain, etc.)

- Symptoms: Healing sign in the shape of a grain of rice on the iris

- Natural Healers: A natural source of vitamins and minerals

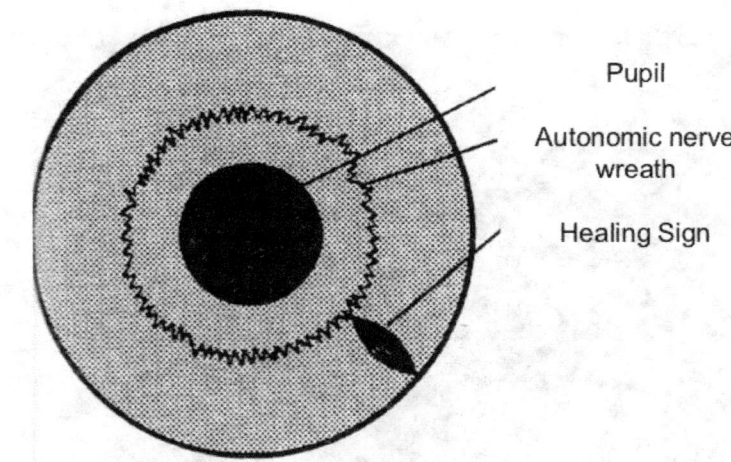

Pupil

Autonomic nerve wreath

Healing Sign

Sample Practice

- Extreme Stress

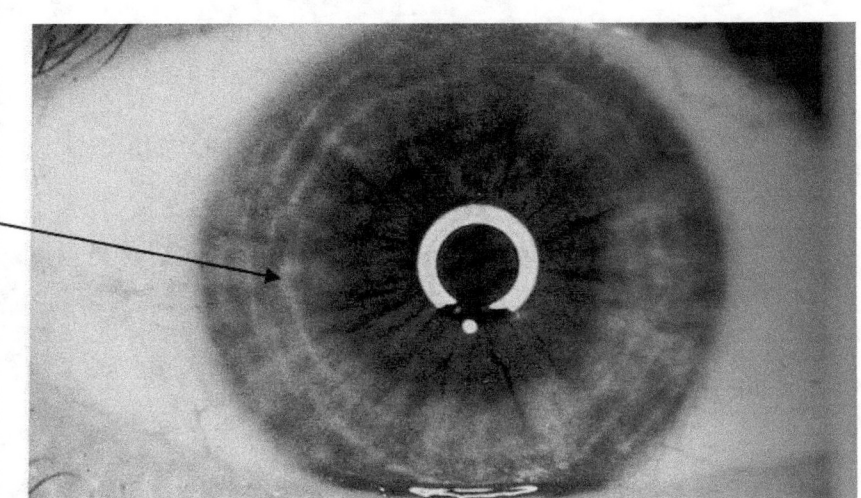

Sample Practice

- Arthritic

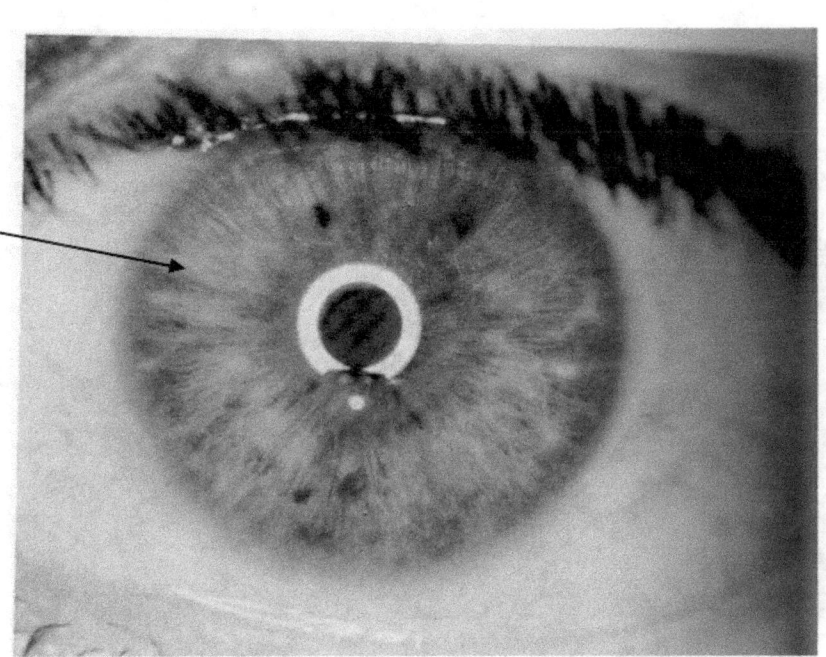

Sample Practice

- Blood Pressure
  Problem

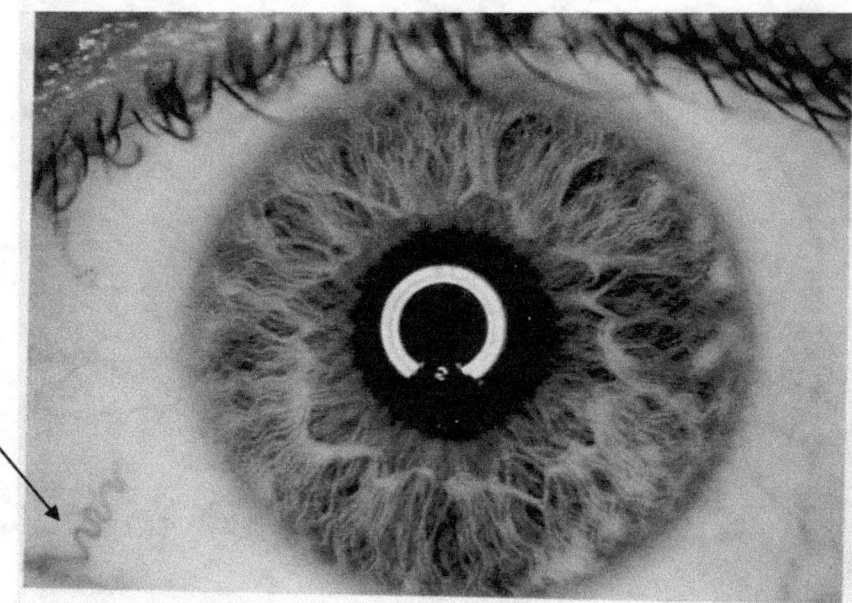

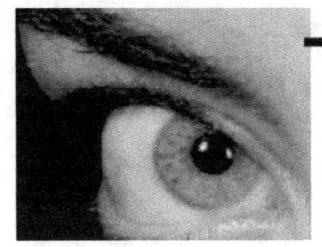# The white of the eyes (sclera)

Should have a crisp, bright white appearance

Reddish scleras: Allergy or calcium, vitamin deficiency

Blue-ish Tint: Poor circulation

Yellow Tint: Lipid/cholesterol plaquing

Brownish Tint: Toxic blood stream (liver or kidney, colon)

Cloudy or milky: Lymphatic problem

Pale/Flat white: Anemia

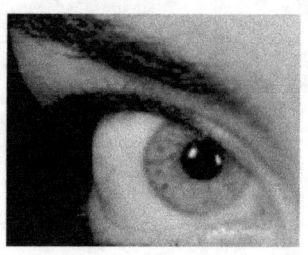

Scleral Signs

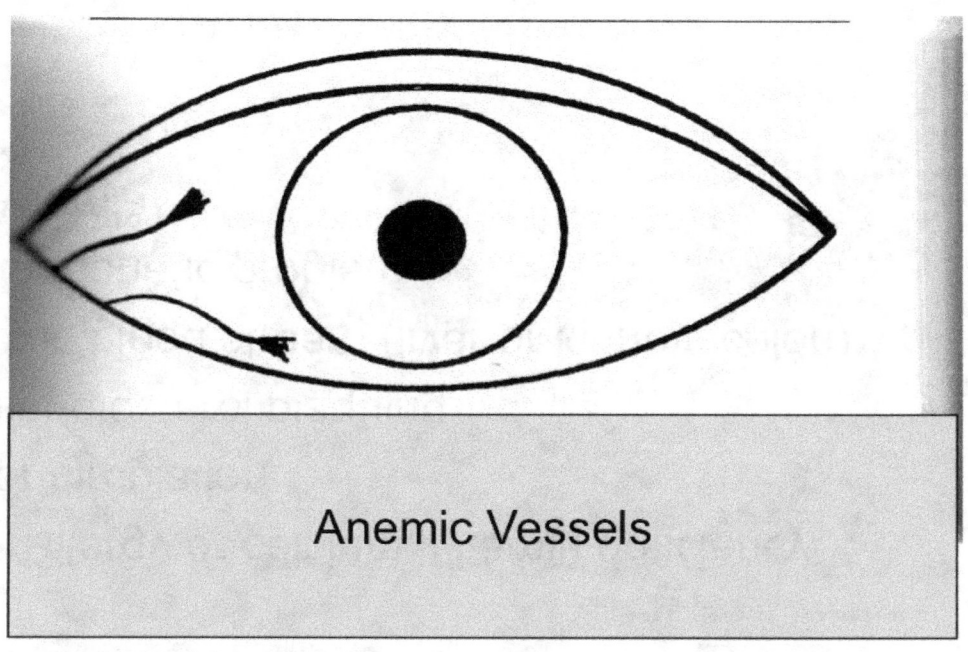

Anemic Vessels

Iron Deficiency and B-Complex

Scleral Signs

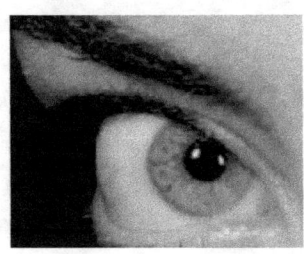

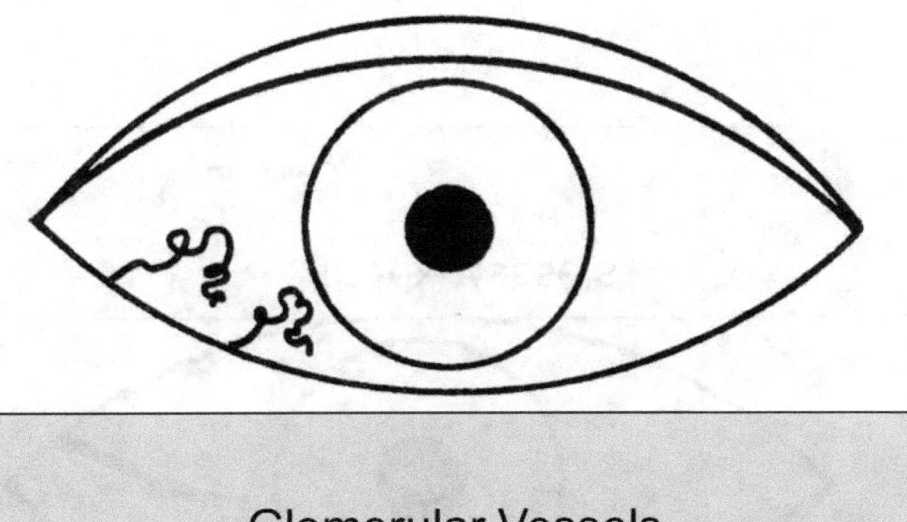

Glomerular Vessels

Hypertension/High Blood Pressure

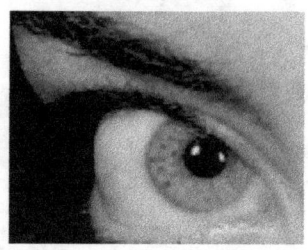

Scleral Signs

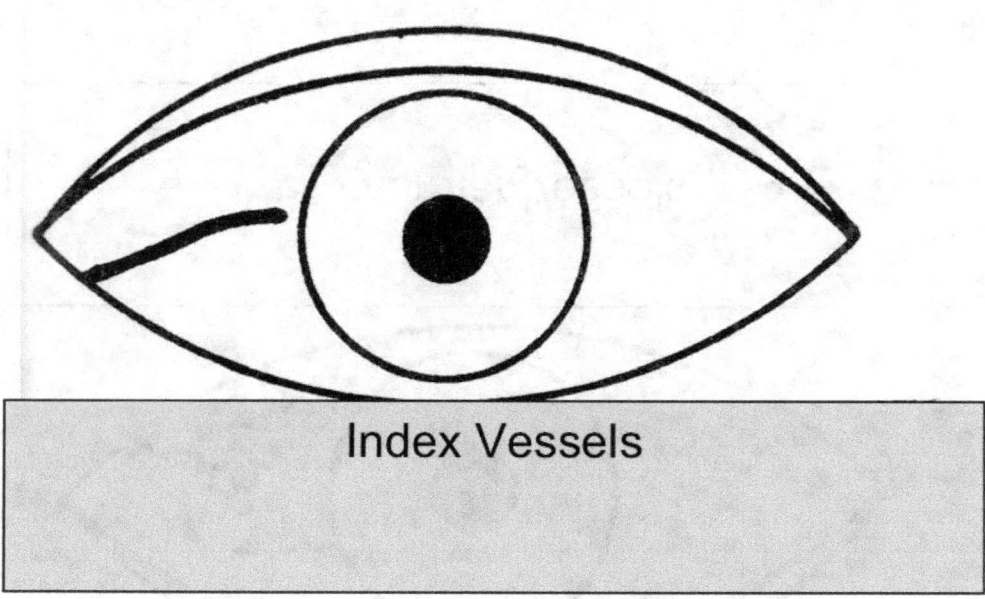

Index Vessels

Thick vessel points to an area in the iris that the body is trying hard to heal. Here: 8 o'clock Right eye = Liver problem

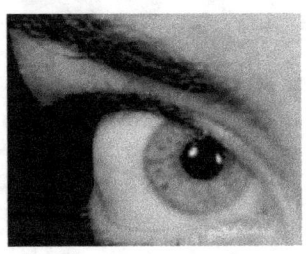

Scleral Signs

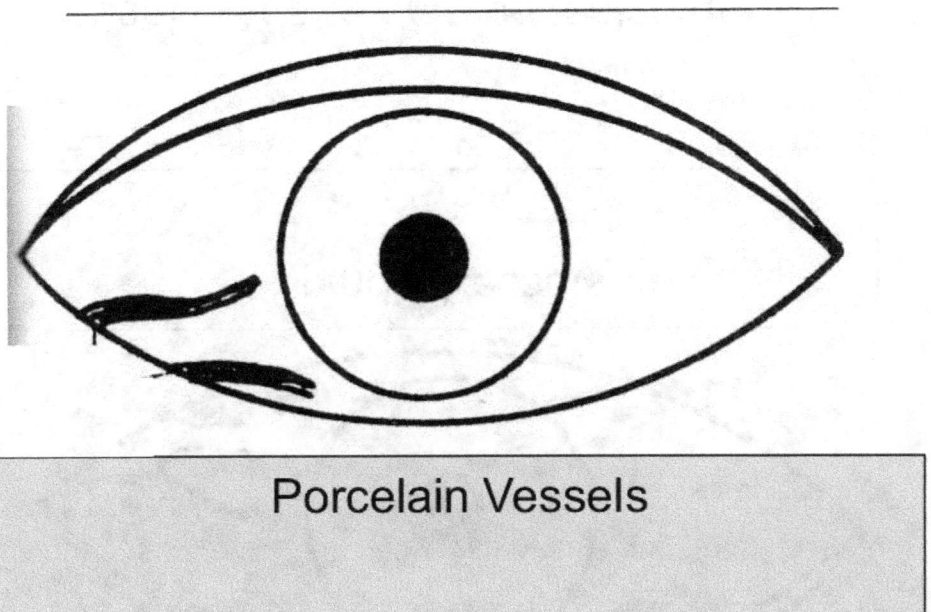

Porcelain Vessels

Atherosclerosis and diabetes (shiny reflective appearance)

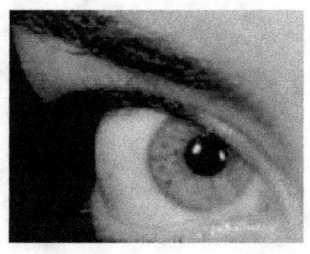

Scleral Signs

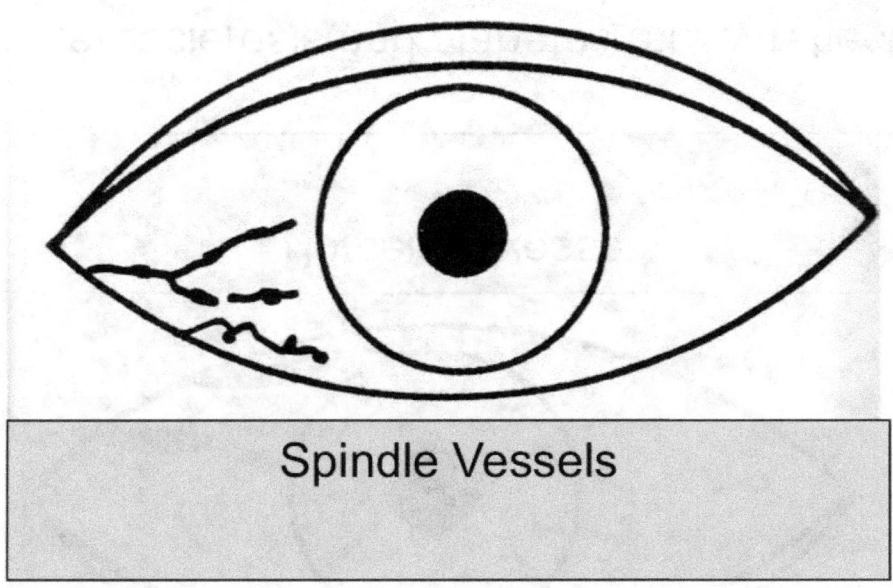

Spindle Vessels

Vessel tends to enlarge it also goes very
constricted. Liver diseases

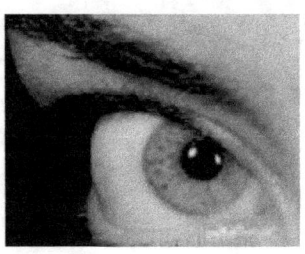

Scleral Signs

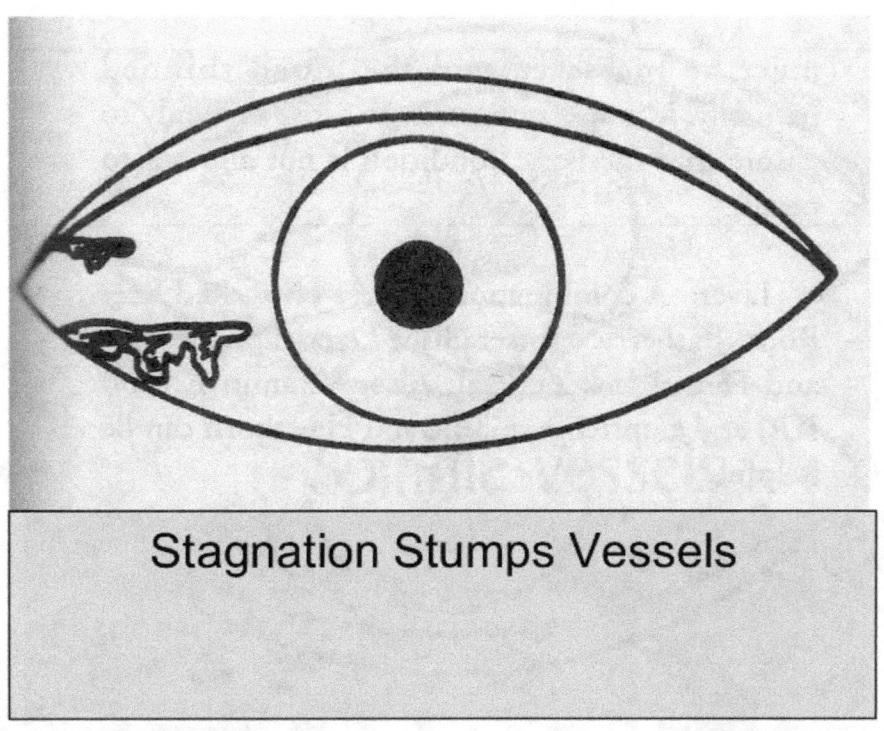

Stagnation Stumps Vessels

Blood clotting. Liver product that may lead to lungs, heart or even brain clots.

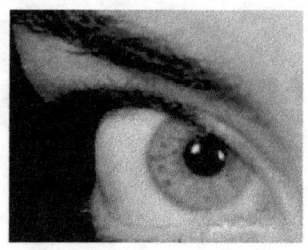

Scleral Signs

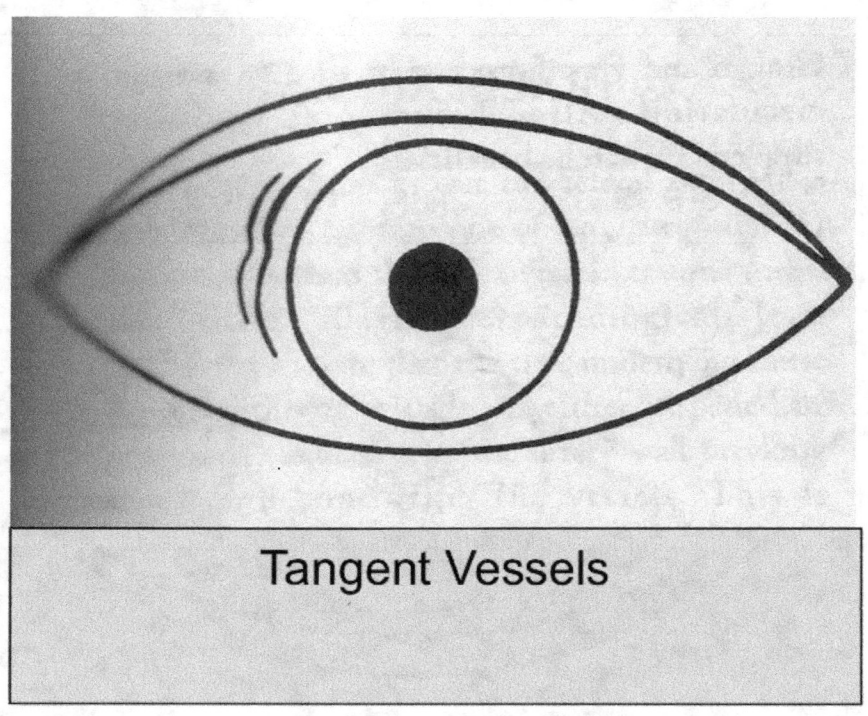

Tangent Vessels

Circulation blockage in body zone

(see Iridology Chart)

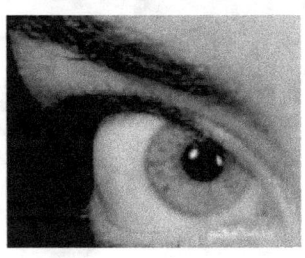

Scleral Signs

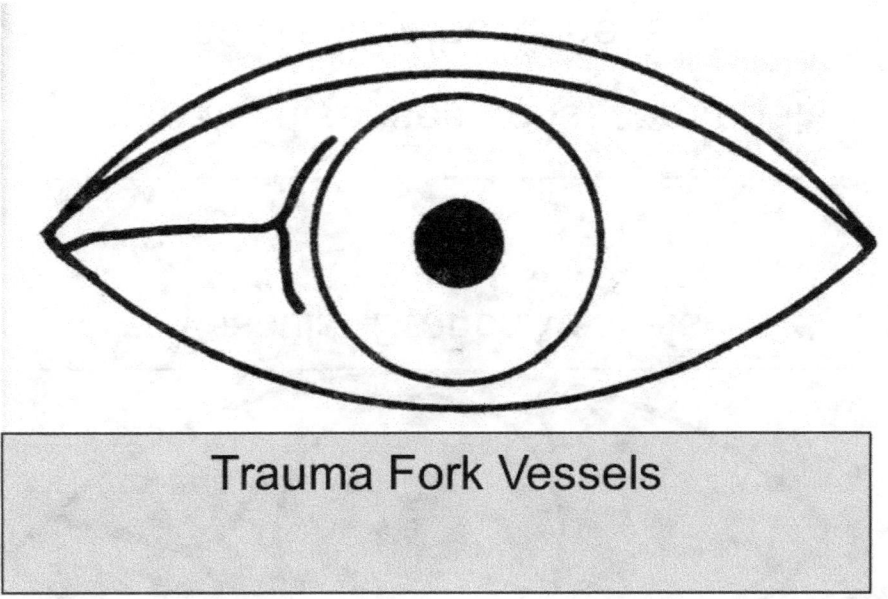

Trauma Fork Vessels

Vessel indicates a trauma/accident of some kind in specific area (see Iridology Chart). It may diminish overtime.

Scleral Signs

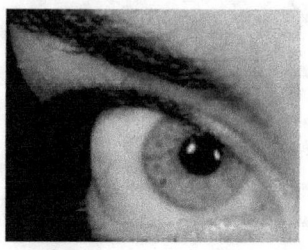

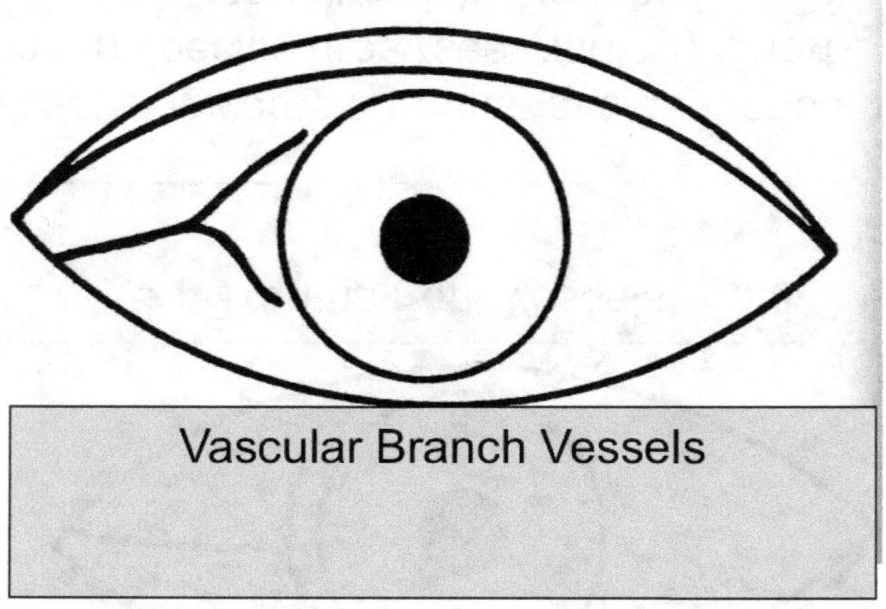

Vascular Branch Vessels

Rheumatic diseases, most commonly

caused by the kidneys

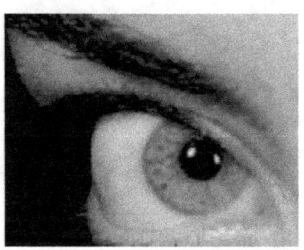

Scleral Signs

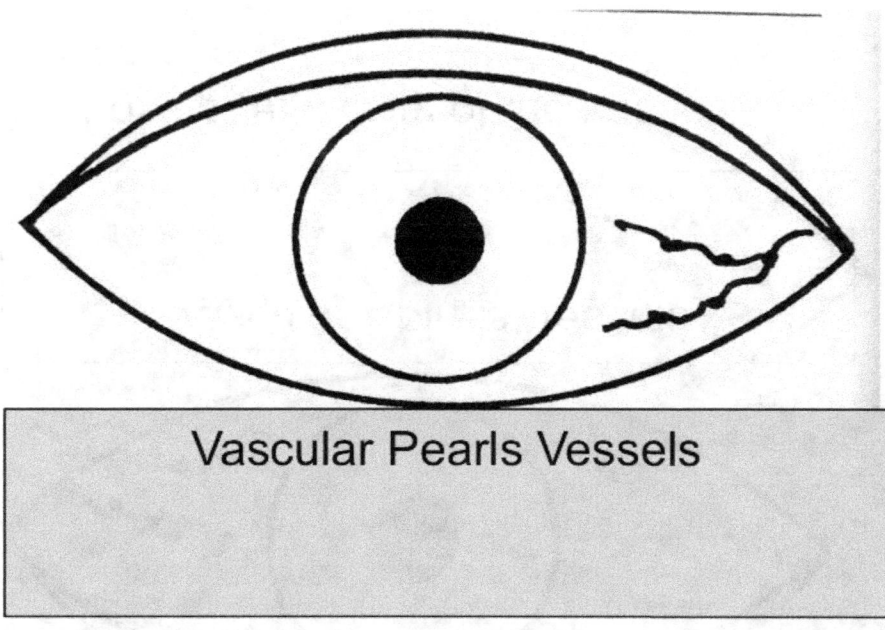

Vascular Pearls Vessels

Weakening of the artery wall (aneurysm)

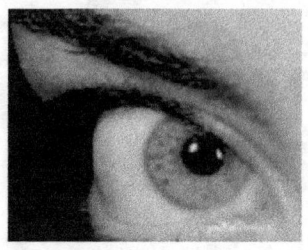

Scleral Signs

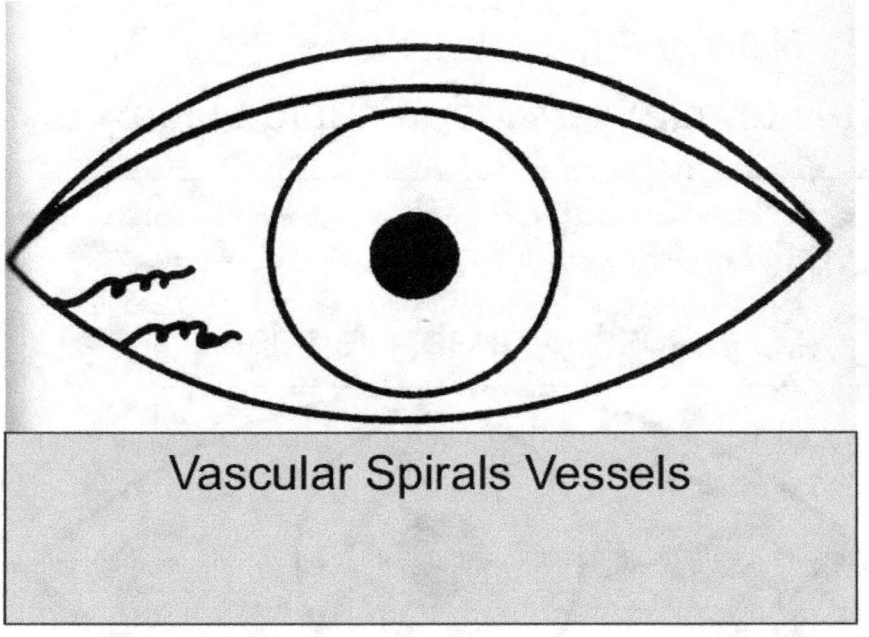

Vascular Spirals Vessels

Loss of elasticity of the vessels

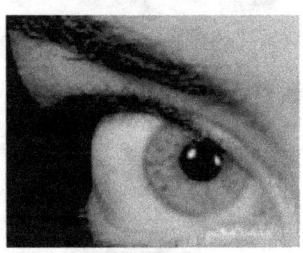

Scleral Signs

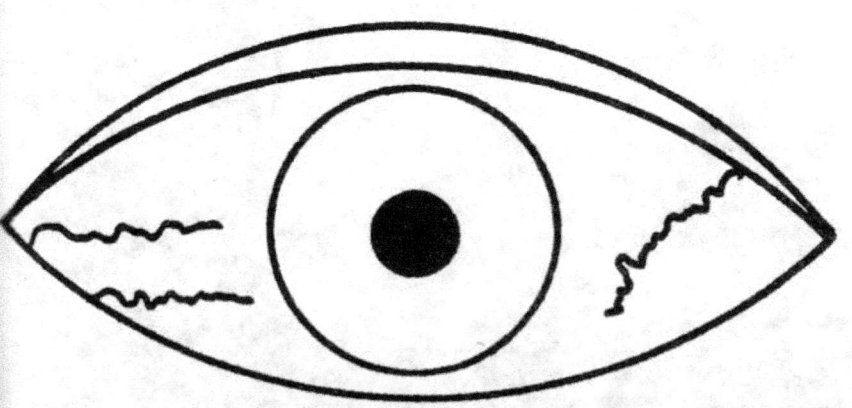

Wandering/Meandering Vessels

Varicose veins, hemorrhoids.
Condition of stagnation/congestion

in the circulatory system

www.ingramcontent.com/pod-product-compliance
Lightning Source LLC
Chambersburg PA
CBHW081309180526
45170CB00007B/2625